Curious Crea

STAR-NOSED MOLES

GAIL TERP

BLACK RABBIT BOOKS

Bolt is published by Black Rabbit Books
P.O. Box 227, Mankato, Minnesota, 56002
www.blackrabbitbooks.com

Marysa Storm, editor; Michael Sellner, designer and photo researcher

Library of Congress Cataloging-in-Publication Data
Names: Terp, Gail, 1951- author.
Title: Star-nosed moles / by Gail Terp.
Description: Mankato, Minnesota : Black Rabbit Books, [2023] |
Series: Bolt. Curious creatures | Includes bibliographical references and index. |
Audience: Ages 8-12 | Audience: Grades 4-6 |
Summary: "Star-nosed moles are animals unlike any other, and curious and reluctant readers alike will be captivated by these exceptional creatures' life cycle, habitats, diet, and threats to survival through carefully leveled text, detailed infographics that emphasize visual literacy, and engaging, colorful photography"– Provided by publisher.
Identifiers: LCCN 2020016580 | ISBN 9781623105686 (hardcover) |
ISBN 9781644664933 (paperback) | ISBN 9781623105747 (ebook)
Subjects: LCSH: Star-nosed mole–Juvenile literature.
Classification: LCC QL737.S76 T47 2023 | DDC 599.35/9-dc23
LC record available at https://lccn.loc.gov/2020016580

Image Credits

Alamy: All Canada Photos, 6–7, 25; Ross Knowlton Nature Photography, 29; Flickr: Hillbraith, 17; Getty: Stan Tekiela Author / Naturalist / Wildlife Photographer, 22–23; iNaturalist: Chunguang Jia, 20–21; Nick Tepper, 26–27; Kuhn Photo: Dwight Kuhn, 1, 3, 4–5, 6, 11, 12, 13, 16–17, 31, 32; PubMed: Duncan Leitch, 18; Science Source: Skip Moody, 11; Shutterstock: Agnieszka Bacal, Cover, 8–9; FedBul, 25; Henrik Larsson, 25; Kseniakrop, 14–15; kzww, 25; Le Do, 25; Nynke van Holten, 25; Vanderbilt University: Kenneth Catania, 22

Contents

CHAPTER 1

A Star-Nosed Mole in Action

A star-nosed mole travels through a dirt tunnel. It uses its star-shaped nose to touch all over the ground. The star never stops moving. It's always feeling for **prey**. In a flash, the mole touches, grabs, and eats a worm. Then it moves on, its star scanning the dirt for its next bite to eat.

How Big Is a Star-Nosed Mole?

TOTAL LENGTH
5.75 TO 8.1 INCHES
(14.6 to 20.6 centimeters)

The Star

A star-nosed mole is nearly blind. But it doesn't need good eyesight to hunt. The star on its nose helps it find prey. The star is made up of 22 short **tentacles**. These tentacles surround the mole's nostrils. They are incredibly **sensitive**.

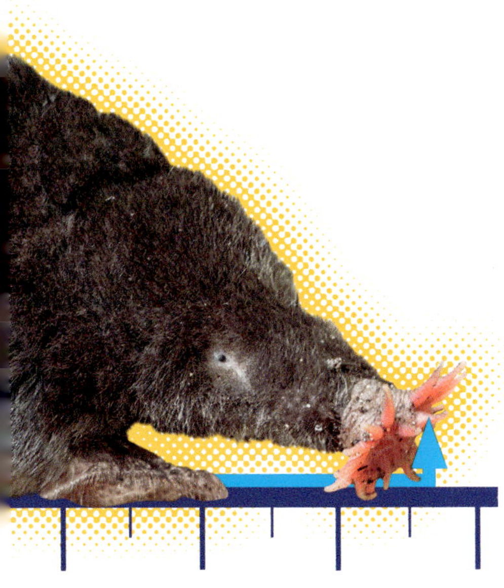

PARTS OF A STAR-NOSED MOLE

FUR
CLAWS

CHAPTER 2

FOOD and Home

Star-nosed moles hunt for prey both on land and in water. On land, they dig tunnels underground. They look for insects and worms. In water, they hunt for small fish. They also hunt for insect **larvae**. Moles are good swimmers. They even swim under the ice in water. These moles are active throughout the year. They hunt both day and night.

Star-nosed moles can smell under water. They blow bubbles. Then they sniff the bubbles back in. If they smell like prey, it's time to eat!

Star-Nosed Mole Habitats

FIELDS

WOODS

SWAMPS

Life in North America

Star-nosed moles are found in the eastern parts of the United States and Canada. They tend to live in wet areas. Their strong claws are perfect for digging tunnels. The tunnels can be near the surface. But they can also be as deep as 24 inches (61 cm). Some tunnels open right into a stream or lake. Star-nosed moles build their nests in the driest parts of the tunnels. They line their nests with grass and leaves.

WHERE STAR-NOSED MOLES LIVE

Star-Nosed Mole Range Map

Europe
Asia
Africa
Australia

CHAPTER 3

The Family

Studying these animals is difficult because they live underground. Experts think they are more social than other moles, though. They believe the moles live in small **colonies**. They might share the tunnels they dig.

Some males and females form pairs in early fall. They stay together until **mating** season in early spring. Females usually give birth in May or June.

The moles store fat in their tails for winter. When food is hard to find, the fat gives them energy.

COMPARING WEIGHTS

Newborns

Each year, females give birth to one **litter**. Their litters have two to seven babies. A newborn has no hair. But its hair soon grows in. It is born with a star around its nose. But the tentacles are folded back. In about two weeks, the star unfolds. The young mole learns to use it.

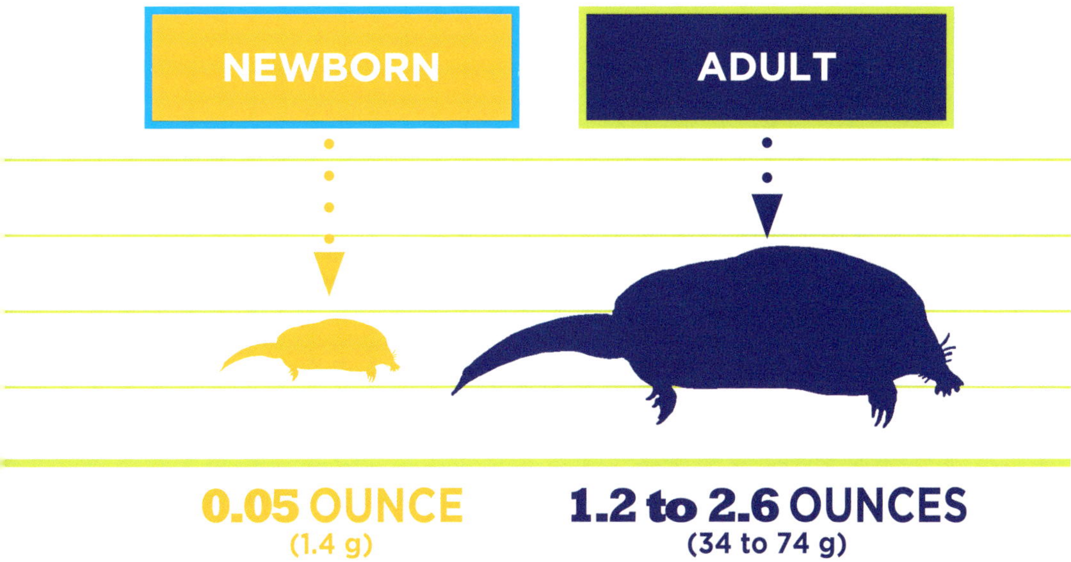

Growing

Newborns cannot take care of themselves. Their mothers must feed them milk. The babies grow quickly, though. When they are about one month old, they leave the nest. They are now independent and can find food for themselves.

By the Numbers

less than
0.25 second

how long it takes a star-nosed mole to find, grab, and eat prey

44

NUMBER OF TEETH

3 to 4 YEARS LIFE SPAN

2.5 to 3.5
INCHES
(6.4 to 9 CM)

TAIL LENGTH

more than
30
seconds

how long a star-nosed mole can stay under water

CHAPTER 4

Danger!

Star-nosed moles have many **predators**. They face danger on land from skunks, foxes, and large birds. Danger in the water comes from large fish. Their strong claws don't help the moles fight predators. They use them only for digging. They can avoid many predators by staying in their tunnels.

Star-Nosed Mole Food Chain

This food chain shows what eats star-nosed moles. It also shows what star-nosed moles eat.

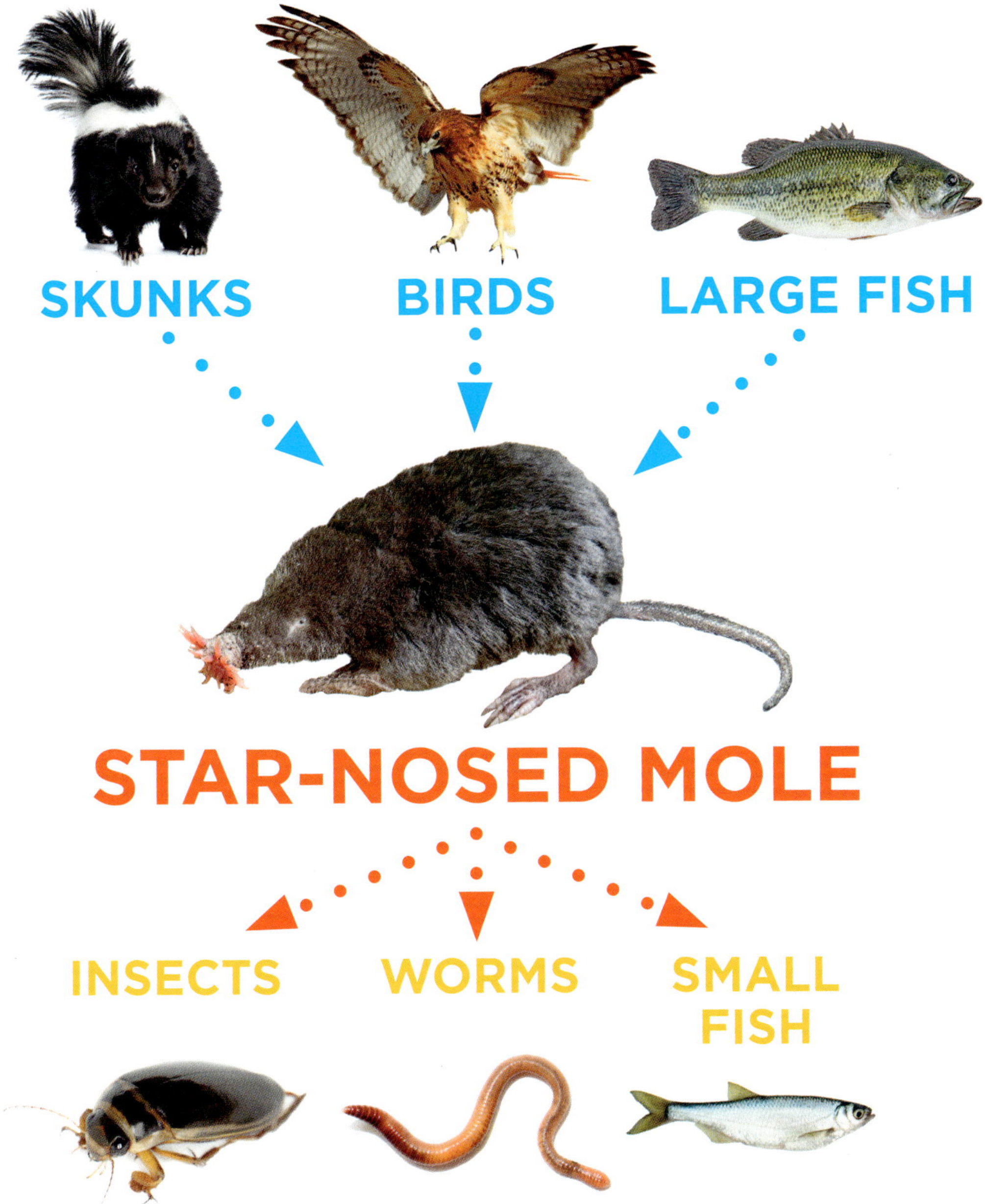

Star-Nosed Moles and Humans

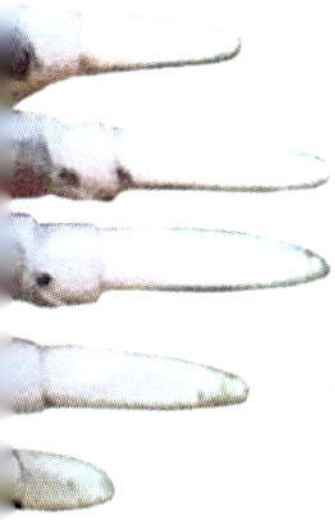

Humans can also be a threat to star-nosed moles. People build on the wetlands where the moles live. The moles can't live on the land and build tunnels if humans take over. They have less land to live on.

Pet cats sometimes hunt star-nosed moles.

Helpful Star-Nosed Moles

People are still learning about these curious creatures. Scientists study the moles' stars and brains. Scientists hope to learn more about how the animals' brains work. Studying moles might help people understand human brains and how people sense things. In time, the work might lead to better treatments for pain.

Star-nosed moles have always been interesting animals. Now they might be helpful too.

colony (KAH-luh-nee)—a group of animals of the same type living closely together

larva (LAR-vuh)—the wormlike form of an animal that hatches from an egg

litter (LIT-ur)—the young born to an animal at a single time

mate (MAYT)—to join together to produce young

predator (PRED-uh-tuhr)—an animal that eats other animals

prey (PRAY)—an animal hunted or killed for food

sensitive (SEN-si-tiv)—quick to detect or respond to slight changes, signals, or influences

tentacle (TEN-tuh-kuhl)—a long, flexible limb

BOOKS

Francis, Alex. *Wild and Wacky Animals: an Alien's Guide.* Early Bird Series. Minneapolis: Lerner Publications, 2021.

London, Martha. *Moles.* Underground Animals. Minneapolis: Cody Koala, an imprint of Pop!, 2021.

Topacio, Francine. *What Do You Know about Underground Animals?* Crazy Creatures of the Animal Kingdom. New York: PowerKids Press, 2021.

WEBSITES

Mole Facts for Kids
www.dkfindout.com/us/animals-and-nature/insect-eaters/moles/

The Speedy, Super-Sensitive Snouts of Star-Nosed Moles
www.ctpublic.org/environment/2020-01-24/the-speedy-super-sensitive-snouts-of-star-nosed-moles

Wacky Weekend: Nifty Noses
kids.nationalgeographic.com/wacky-weekend/article/nifty-noses

INDEX